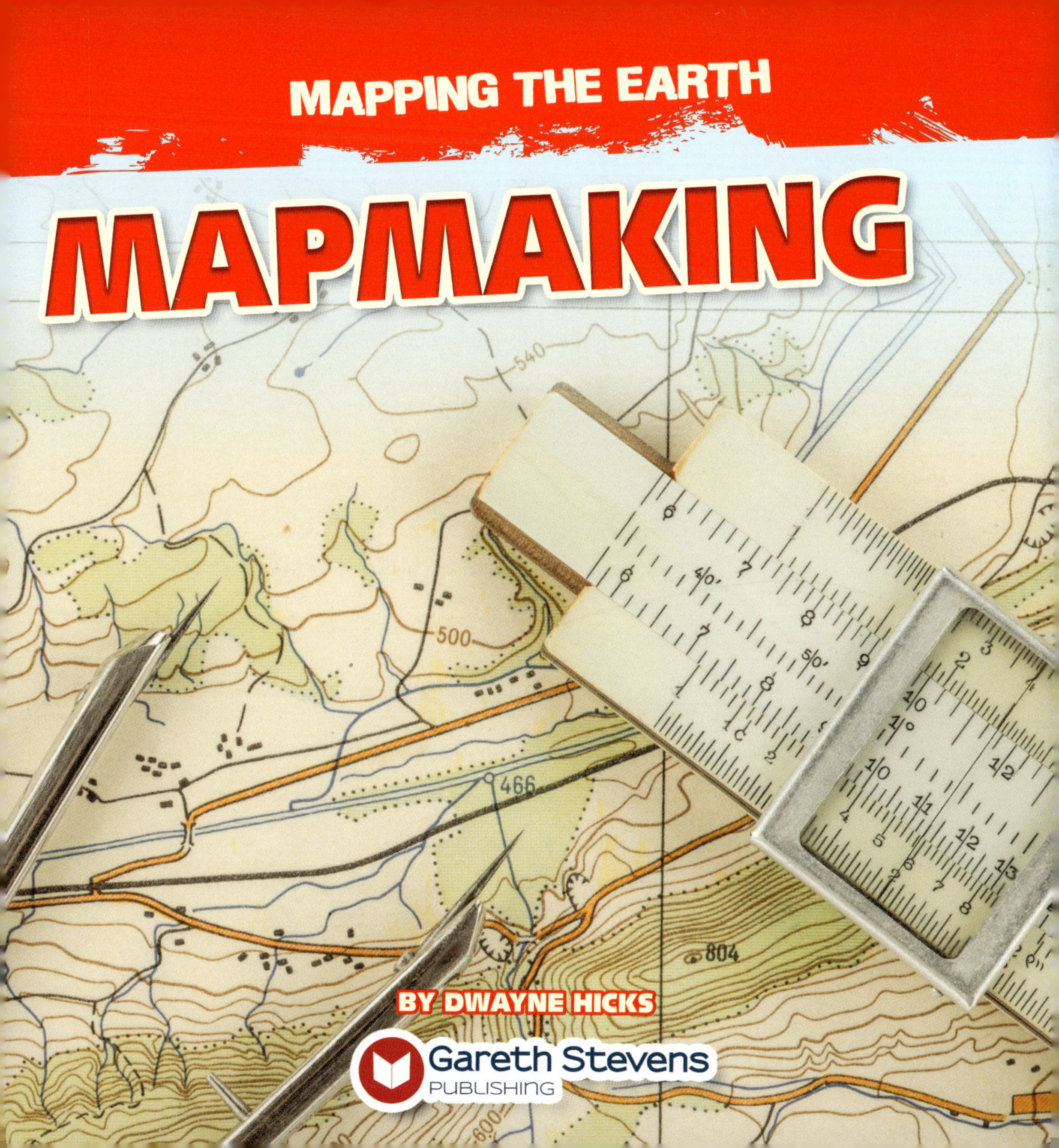
MAPPING THE EARTH
MAPMAKING
BY DWAYNE HICKS
Gareth Stevens
PUBLISHING

Please visit our website, www.garethstevens.com. For a free color catalog of all our high-quality books, call toll free 1-800-542-2595 or fax 1-877-542-2596.

Cataloging-in-Publication Data
Names: Hicks, Dwayne.
Title: Mapmaking / Dwayne Hicks.
Description: New York : Gareth Stevens Publishing, 2023. | Series: Mapping the Earth | Includes glossary and index.
Identifiers: ISBN 9781538278031 (pbk.) | ISBN 9781538278055 (library bound) | ISBN 9781538278048 (6pack) | ISBN 9781538278062 (ebook)
Subjects: LCSH: Maps–Juvenile literature. | Map drawing–Juvenile literature.
Classification: LCC GA130.H53 2023 | DDC 526–dc23

Published in 2023 by
Gareth Stevens Publishing
29 E. 21st Street
New York, NY 10010

Portions of this work were originally authored by Todd Bluthenthal and published as *Making Maps*. All new material in this edition authored by Dwayne Hicks.

Designer: Tanya Dellaccio
Editor: Kristen Nelson

Photo credits: Cover Slava17/Shutterstock.com; pp. 2-24 (background texture) Triff/Shutterstock.com; p. 5 MNStudio/Shutterstock.com; p. 7 Peter Hermes Furian/Shutterstock.com; p. 9 (background) The_Pixel/Shutterstock.com; p. 11 Wikrom Kitsamritchai/Shutterstock.com; p. 13 Mariusz Szczygiel/Shutterstock.com; p. 15 Alex Oakenman/Shutterstock.com; p. 17 Michael Gordon/Shutterstock.com; p. 19 Odua Images/Shutterstock.com; p. 21 akvarelmed/Shutterstock.com.

Printed in the United States of America

CPSIA compliance information: Batch #CSGS23: For further information contact Gareth Stevens, New York, New York at 1-800-542-2595.

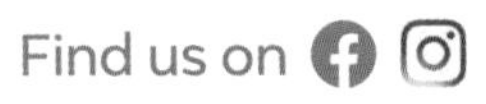

CONTENTS

Boldface words appear in the glossary.

So Many Maps!

Maps can show a lot of **information** about a place. We have maps of towns and cities. We have maps of the world too! The people who make maps are called cartographers. They make maps so we can learn about places.

What Is a Scale?

Maps use scales. A scale shows what a **measurement** on the map means in **distance** on Earth. When making a map, the paper will be much smaller than the place you're mapping. Using a scale fixes this problem for cartographers.

ARCTIC OCEAN
USSIA
ALASKA
(U.S.)
CANADA
ERING SEA
PACIFIC OCEAN

SCALE →

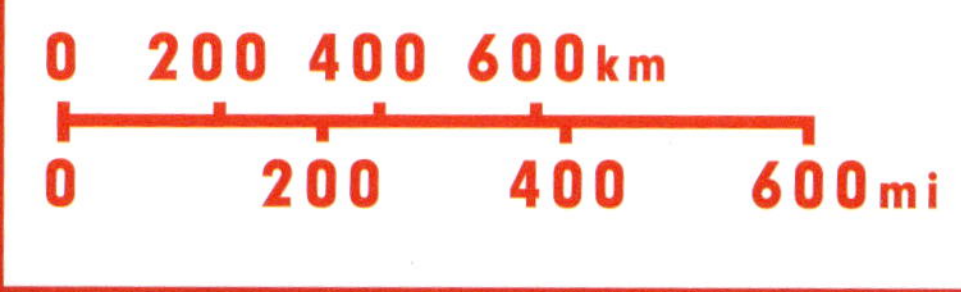
0 200 400 600km
0 200 400 600mi

Sizes on the Map

Cartographers use math to make scales. But you don't have to when you make your map! Just remember the size of things in real life when you're drawing them on your map. For example, draw a house much larger than a person.

TO SCALE
NOT TO SCALE

Which Way?

A compass rose is a **symbol** cartographers use to help people read maps. A compass rose shows the cardinal directions—north, south, east, and west. You can use a compass rose to tell readers which direction they need to go.

COMPASS ROSE
NORTH
WEST
EAST
SOUTH

Bird's-Eye View

When making a map, pretend that you are looking down from the sky like a bird. Most maps show a place from above. This is often called a "bird's-eye view." Using this **perspective** allows readers to see a whole city!

13

The Map Key

Everything on a map has a color, symbol, or shape. Water is often shown in blue. Parks are often shown in green. Roads are black lines. Houses can be blue squares. Maps use keys to show what the different shapes and symbols mean.

KEY
ROAD
PARK
HOUSE
WATER

Make Your Own Map

You can draw a map of your neighborhood. Go outside and take a look around. Are you standing on a lawn or a sidewalk? What's to your left and right? Think about how far your home is from other buildings.

Gather Information

Take a walk down your street with an adult. Gather information as you go and write down what you see. Point out parks, roads, friends' houses, trees, street signs, and even dogs! You should also carry a **compass** so you know which way is north.

Time to Draw!

The first thing you can draw on your map is your street. Draw each house on your street in the order they are on the street, but from a bird's-eye view! Then you can add colors, shapes, a key, and a compass rose.

GLOSSARY

compass: a tool for finding directions

distance: the amount of space between two places

information: facts

measurement: a size, length, or amount found by measuring something

perspective: a way of showing how close or far places are that matches the facts in the real world

symbol: a picture or shape that stands for something else

FOR MORE INFORMATION

BOOKS

Boyer, Crispin. *National Geographic Kids Ultimate U.S. Road Trip Atlas*. Washington, D.C.: National Geographic Kids, 2020.

National Kids. *Beginner's United States Atlas*. Washington, D.C.: National Geographic Kids, 2020.

WEBSITES

Maps and Mapmaking

online.kidsdiscover.com/discover/maps?ReturnUrl=/discover/maps

Learn more about maps at this interactive website.

Maps for Kids

www.maps4kids.com

Find links to a variety of maps and information related to different areas of the world.

Publisher's note to educators and parents: Our editors have carefully reviewed these websites to ensure that they are suitable for students. Many websites change frequently, however, and we cannot guarantee that a site's future contents will continue to meet our high standards of quality and educational value. Be advised that students should be closely supervised whenever they access the internet.

INDEX